A Practical Guide to Producing Heather Honey

Tony Jefferson

Northern Bee Books

A Practical Guide To Producing Heather Honey
© 2014 Tony Jefferson

ISBN 978-1-908904-61-4

Published by Northern Bee Books, 2014
Scout Bottom Farm
Mytholmroyd
Hebden Bridge
HX7 5JS (UK)

Front cover: © John Phipps

Design and artwork
D&P Design and Print
Worcestershire

Printed by Lightning Source, UK

A Practical Guide To Producing Heather Honey

Tony Jefferson

Northern Bee Books

Contents

I ACKNOWLEDGEMENT

I suppose it is fairly obvious that this book would not be possible without the support of two important influences in my life. All the information herein is based on information passed down to me over the years by my father, Allan. What I have attempted to do when producing this, was to record what he has taught me over the years. If there appears to be gaps, I make no apology – it's just that he never taught me those bits, so they cannot be that important!

My wife, Anne, is also responsible for helping me to produce this book, far more than she realises, but in an indirect way. If anyone has ever tried to call me on the land line over the beekeeping season, she usually has to reply *"Please call him on his mobile, as I have no idea where he is, or when he will be home"*.

Beekeeping never really goes to plan and I am sure that many of you can relate to this. Especially so if you are still looking for that illusive queen in order to finish the artificial swarm that you started four hours ago and are still muttering *"She has to be here somewhere!"*

The Jeffersons

Left to Right: Tony, Richard and Allan

2 INTRODUCTION

Sometime ago I wrote a short article on heather honey for BBKA News. Following many favourable comments regarding the article, I was asked if it would be possible to extend the article into a short and simple book on how to obtain a decent heather crop. This work was to cover important information on breeding bees, and the organisation and practicalities of managing colonies for those who take their bees to the heather.

I hope therefore that you will find the following information useful in your own beekeeping. I have attempted to keep the techniques as simple as possible - all of which have been tried and tested in our own beekeeping practices.

If you have not seen my other short book "The Jefferson Beekeeping Guide" then it is worth getting hold of a copy, as many of the concepts, outlined here, are covered in the book.

Those of you who follow my advice will undoubtedly increase their heather crops significantly. However, there will be some aspects of my beekeeping advice which will go against the conventional or establishment viewpoint – I make no apologies for this - apart from replying *"Well, who is correct ? I know I am!"*

To give you a head start the main controversial areas are:

▶ absence of queen excluders
▶ glass quilts on every hive
▶ use of minimum brood and a half
▶ single brood in WBC hive for winter
▶ hive insulation all year round
▶ flat floors
▶ black/native bees.

The three generations of the Jefferson family have based our beekeeping on an annual cycle of activities leading up to the anticipation of two weeks of decent weather in August. It may seem a great amount of effort in the hope of a good weather window – but we would not be organising our activities and working for fifty weeks if results didn't justify the work involved.

By operating as we do - concentrating on breeding lots of bees early in the season and drawing out wax - we find our other honey yields are improved and that swarming is reduced as if by accident.

Heather honey is a premium quality honey and it is not too difficult to ensure that everything humanly possible comes into place at the right time. A fair amount of praying for good weather is also worthwhile!

Apart from the weather, everything else is down to the skills and planning of the beekeeper to get the colonies ready in the given timescale. The concepts outlined here may appear difficult to comprehend, at first, but in practice they are not actually too difficult - as long as a good understanding of the individual bee and the colony life cycle is known and understood. Understanding the natural history of the colony and the timing of key events is vital to good beekeeping and must be clearly understood, I do not intend to go into too much detail about this just now, but the basics will be covered where required.

The information and advice here is based on the use of British Standard- sized National frames, both normal deep and shallow sizes and thus covers National, Smith and WBC types of hive.

Our floorboards for heather going

top view

underside

funnel entrance

3 TYPES OF HEATHER HONEY

There are basically two types of heather honey:

▸ Bell heather - (my personal favourite) which flowers from July onwards, in smaller patches. It is a dark-coloured, ruby red, liquid honey which can be extracted normally.

▸ Ling heather - which flowers later from August onwards and which is found covering vast areas. The honey colour can range from dark brown to ruby red. It has the characteristic small air bubbles and is thixotropic. It has some minor difficulties with getting it from the comb to jar.

Bell Heather (*Erica cineraria*)

Ling Heather (*Calluna vulgaris*)

The vast majority of heather honey found on sale in the shops is not actually pure heather, it is blended to some extent, often by the bees themselves or by the beekeeper, it is therefore quite difficult to compare various heather honey samples.

To produce really first class (or pure) ling heather - which has the highest value of all our native honeys - does take a little preparation to ensure there are no traces of flower honey mixed in with it. To be honest, it is nearly impossible to ensure that there is no other nectar, but steps to minimise it can be taken.

For exhibition or show bench purposes, the honey should be as much pure heather as possible; it should be displayed in the gel state with a uniform size of bubbles. I have actually had honey excluded in major national shows for the bubbles being too large - as well as them being too small! To me this just indicates that the judges are not experienced enough to deal with ling heather honey.

Pure ling heather will not granulate for some considerable time, as many other honeys do, but even a trace amount of other honey will 'seed' it, which will make it granulate prematurely.

One major disadvantage is that ling heather does have a higher water content than other honey – so it is more liable to ferment. To reduce this risk, refractometers can be used to measure the water composition of the sample of honey, but many beekeepers do not have such equipment. A more simple method is to ensure any stored honey must come from combs that were properly capped prior to them being removed from the hive. If you do have some unsealed honey, make sure it is used rapidly and not stored, or make some mead with it.

4 HEATHER STANCE

The location that hives are taken to on the moors is often referred to as a "heather stance". Finding a good location does take a considerable amount of time and, once found, is worth preserving. What has to be remembered is that every piece of land in the country is owned by somebody, somewhere. It is not acceptable just to drop hives off on the moor without the correct permissions. Unfortunately this does seem to happen from time to time with insensitive migrant beekeepers. This irresponsible behaviour causes a great deal of problems for beekeepers in the neighbourhood, the association often becomes involved, and results in a loss of goodwill towards local beekeepers.

A well-managed and maintained grouse moor is the key requirement for your chosen site. The area will have been managed in such a way that plenty of young heather shoots are made available. The primary purpose of the young shoots is for the grouse to feed upon and are formed after the heather has been previously burnt off. It is these young shoots that provide the best flowers that yield the nectar, the older or long, leggy heather plants will produce very little. Areas of the moors can vary considerably, so it is useful to have a series of stances available, if you run a lot of colonies.

This heather moor has been carefully managed with the plants at the right height for good flower production. Strips of lank heather are usually burnt in February and the new growth will provide grazing for sheep as well as young shoots for grouse. (John Phipps)

When placing bees on the moors, advance reconnaissance is vital to ensure the hives do not pose any problems to the gamekeepers and beaters as well as others who may use the land such as walkers and picnickers. Studying the area carefully to ensure that there is shelter from wind and that there is no chance of flooding are important considerations. Other potential problems which must be taken into account when the hives are on the moors, include the presence of sheep, the possibility of vandalism, whether the moor is inhabited by snakes, and whether the bees are at risk of getting 'run over' by cars.

It is best to position each hive directly on the ground (without hive stands), as this will prevent them from being knocked over by sheep which love to scratch themselves on the sides of the hives. It is best to leave the hive straps in place to prevent the boxes being dislodged. Also, place some thin plywood in front of each hive to prevent grass growing over the entrance and space them at a minimum of two sheep distance apart.

Hives in place on the moor with straps in left around them and boards in front of the hive to prevent grass growing up and blocking the entrance. *(Colin Weightman)*

When taking the hives off the moor snakes, particularly grass snakes, are often found curled up under the hives where it is nice and warm and dry for them. So being aware of this, take care when picking up hives at the end of the flow, and be sure to wear gloves.

Grass snake (*Natrix natrix*). *(Jeff de Longe)*

I often see hives positioned in poor locations far too close to main roads. The bees have to fly over the roads to forage on the moorland usually flying in all directions. When returning, loaded with nectar, they fly much lower. If they have to fly over a busy road,there can be many casualties as many are hit by passing traffic.

Ideally a sheltered location should be sought, one that gives some protection from the prevailing wind. Generally the hives are best facing in a southerly direction, but this may not always be possible.

Hives on the moor should be placed directly on the ground, spaced at least two sheep widths apart, ideally facing south, and with shelter from the cold winds. (*John Phipps*)

Places to avoid are low-lying areas which are prone to flooding. Thunderstorms and flash floods are not uncommon events in August and September.

An area that provides easy access for vehicles is really important. Should there be a good honey crop any removed supers need to be quickly taken away and into the security of an place that minimises robbing - being covered up and put in the car boot is best.

Transporting the hives to and from the moors needs to be as simple as possible and carrying hives is an activity that is best to be avoided.

5 COLONY SELECTION & DEVELOPMENT

To put this simply, the basic requirement is to have:
- ▶ the right equipment,
- ▶ the right bees,
- ▶ at the right age,
- ▶ at the right time.

This applies equally for maximising any type of honey crop and the dates / timings explained later can be applied to any particular honey crop and varied dependent on the area in which you keep your bees.

All my beekeeping activity is totally devoted to maintaining good productive colonies of our local Dark Bee – *Apis mellifera mellifera*. A great deal of effort each year goes into the positive selection and breeding out of undesirable characteristics. I am fortunate to live in an area where the dark bee predominates, but from time to time bees with yellowish markings do seem to occur, especially in years when good weather for matings occur. Yellow-banded bees are not immediately culled, but they are only kept for a maximum of the same season – steps are taken to eliminate them before going into winter, the colonies are retained as a resource for honey production.

There are a number of reasons for propagating the dark bee and many will argue either way. But, as my father has been doing this with good results for around 70 years, there is really little point in me experimenting with anything else, for a number of reasons:
- ▶ The results we get are favourable.
- ▶ The bees are not overly prolific but do build up to significant numbers when required.
- ▶ I may as well use Dad's 70 years of experience rather than start from scratch as I would not see the results during my life time!

The native black bee I believe to be well-suited to the harsh moorland environment and the periods of sustained hard work on the heather moors that is demanded to get a good heather crop. Even in August the ambient daytime temperature variation can be considerable, but the night-time temperatures can be down to near frost points. Due to the dark bee being particular productive at low temperatures, they are off out foraging early in the day and still remain working into the early evening. If the weather deteriorates, long periods of confinement can occur which reduces the egg-laying rate of the queen, but the colonies do show a recovery when taken off the moors.

Every year we have the same debate *"When should we aim to have all the hives on the moor?"* Times do vary but it is anticipated that the heather should yield nectar around 'the

Glorious 12th' (the traditional start of the grouse shooting season). It is best therefore to use the 12th of August as the date for all hives to be resident on the moor. Seasonal variations do occur but generally only by a week or two.

My Dad is quite famous for some of his quotes, one regarding moving bees is *"We are not taking them to the moor for a holiday"*, which basically means there is little point in taking small hives to the heather, they need to be productive colonies or we have wasted our time and effort. The aim is therefore, to have colonies with a minimum of 9-10 frames of brood, in the brood chamber. (This is based on BS National frames.) The more productive colonies (I will refer to them as 'production' colonies) which are ideal for heather production will be built up to be on a brood box and a shallow brood extension (commonly known as a "brood and a half").

I am firmly of the belief that a brood and a half together with NO queen excluders is the best combination for heather honey production. I look forward to the many arguments that the above statement might yield - but those who know the "Jefferson Way" will realise that we are able to vigorously argue the benefits of this method.

I find the whole issue of using queen excluders without thinking about their purpose as somewhat strange and bewildering. I just cannot understand why a beekeeper choses to restrict a queen from laying her eggs. I prefer to call them "queen restrictors" or "honey excluders" - then the whole understanding of them takes on a different meaning. I have yet to see a feral colony, in a tree for example, where there is a fixed mechanical queen excluder present, so we need to consider these 'man-made' contraptions carefully before using them.

Queen restrictors do actually exist in a natural colony, they are formed by an arch of sealed honey, as a queen will not pass over sealed honey to lay eggs. This is one of the reasons why we 'under-super' to maximise brood production. Under-supering means placing supers, with empty drawn-out combs, under the sealed honey to maximise brood rearing and honey collection. Many beekeepers are well aware of the need to add space for honey storage, but they do not automatically consider space for brood production.

The primary aim to maximise heather honey is to have the maximum quantity of foraging bees available for the whole period the bees are on the moors. This means ensuring plenty of eggs are laid six weeks prior to the beginning of August, ie from mid-June and through the whole of July.

Unfortunately this period falls right in the middle of the usual 'June Gap' when the amount of incoming nectar reduces and the queen's egg-laying rate falls accordingly. Stimulative feeding is therefore required to keep the queen in lay - a contact feeder with 1:1 syrup being recommended. Do make sure a rapid feeder type is not used as you need the bees to use the syrup immediately from the contact feeder - and not store it as they would do with a rapid feeder. It is equally important at this time to provide plenty of comb space for the queen to lay in.

Another problem is that after the longest day (21st June) queens naturally reduce their egg-laying rate as they commence their colony reduction for the winter period, so they need to be encouraged to lay eggs during this important phase. Non-native bees do not follow the same pattern, but that results in too much brood and therefore too much food demand as the colonies approach the winter time.

To achieve decent 'production colonies' it may be a requirement to unite small colonies so that they fill a brood chamber thus providing a plentiful supply of foraging bees. By uniting I mean using the newspaper method a description of which follows in later sections of this book.

Early young queens' colonies often do well on the moors provided they are encouraged to build up quickly when the new queen is laying. For example, a young queen produced in May will have time to build up in colony strength for August, but only if in the early stages of the colony growth some frames of emerging bees from another hive is given. This increases the availability of young nurse bees which will boost colony development. Colonies with later, or July queens, although the queens may be mated well, do not have the time to build up sufficiently, but can be good producers the following year.

We operate on a policy of never destroying a good queen cell, for should a queen emerge successfully and mate adequately, then the brood and young bees can be united to colonies thus supplying them with more bees. Simply consider some simple mathematics: if three newly-mated queens are each laying 1500 – 2000 eggs a day in their first year, this is obviously much better than a single second-year queen laying typically 2000 a day. Selecting the best new queens and uniting them to heather colonies is fairly straightforward. But consider keeping the other removed queens in 'insurance' nuclei in case of loss or damage to queens on the moors.

6 ARRANGING BROOD PRIOR TO MOVING

One key aspect to increase the honey crop, is the concept of manipulation of brood frames, this is not practised by many beekeepers. The intention is to maximise queen performance and hence increase brood production. I will take this step by step as it is of major importance:

Starting point – NO queen excluders:

▶ Prior and up to taking them to the heather, the task required is to get good heather gathering colonies, with plenty of brood and plenty of foraging bees.

▶ In early June and onwards, on a weekly basis, during swarm inspections, spread the centre of the brood nest and place at least one (or more if a strong colony) of empty drawn comb(s) in the centre to ensure the queen lays lots of eggs in the middle of the brood nest. This applies both to the brood nest itself and any brood extensions that are used.

▶ By doing this, you are attempting to maintain a pyramid of new eggs and larva with the old brood on the outside as you do not want the queen to be short of available cells for egg-laying.

▶ If any combs contain an arch of sealed honey – this will keep the queen below the sealed honey and she will reduce laying. Either move these combs up into the super - or bruise up the sealed cells with the hive tool and then the bees will move the honey upwards themselves. Remember - sealed honey is nature's queen excluder!

▶ The old brood which is displaced to the outside will also ensure that any honey is taken up and stored in the supers

▶ At this stage, producing eggs is the major consideration and since it often occurs during the traditional 'June Gap' stimulative feed should be applied and continue; however, there is likely to be plenty of pollen available during the June gap if not nectar.

▶ Whilst the colonies are on the moors there is also a need to encourage egg-laying as the new bees will become the 'winter bees'.

The next step is very important:

▶ The week prior to moving the colonies to the heather stance, the young brood is moved to the outside (reverse of above). The older brood is now in the centre and this will emerge first whilst on the moor. The queen will then re-lay in these cells to produce 'winter' bees. The young brood on the outside makes sure the honey goes in the top. The reason for not using queen excluders is because the queen's laying rate will already be reducing at this time of year and every egg produced is likely to be needed for winter bees.

▸ Prior to moving colonies, any sealed summer honey should be removed. If any is unsealed it can be placed on other colonies or left on the top of the hives that are going to the moor. Consider very carefully that if you do remove all the summer honey, if the weather changes when the bees are taken to the moor, they may well become short of food in a very short time period (about 24 hours) – the need for feeding is not unheard of in August!

▸ Put supers with drawn-out combs on each hive prior to moving as plenty of space for the bees to expand in transit is required. A general rule is to add one super for every box of brood - so if the colony consists of a brood and a half, then two empty supers are required.

▸ A main advantage for honey production is to have plenty of drawn-combs available and not have to rely on the bees using heather honey to draw out wax. Remember it takes at least 8 lb (some say 10 lb) of honey to produce 1 lb of beeswax. Consider a super of 11 frames, approximately, this would be about 1 lb of beeswax to produce all the drawn out cells so that's immediately 8-10 lb of heather honey 'lost'.

Deep frame with heather honey - the comb being drawn from foundation will use a lot of the incoming nectar.

- Plan ahead and consider getting plenty of drawn combs from earlier crops, spin the honey out and give them back to the bees for clearing up any residues. Many people use thin super or even starter strips - these methods will lose even more time for honey storing as well as a loss in the total amount of valuable heather honey to be harvested.
- Maintaining levels of top insulation whilst the hives are on the moor is also important, due to the cold night-time temperatures. It takes far less energy for the bees to ventilate and cool the hives than it does to keep them warm. The valuable heather nectar will be ripened, turned into honey and sealed much quicker if the hive temperature is constant and the bees only have to ventilate, so top insulation must be provided.
- Our floor boards are our own version of the old Steel and Brodie 'heather' floor, these are flat floors that have an internal funnel-type entrance with the bees entering from beneath the centre of the brood box. This has major advantages as the funnel acts as an internal alighting board, so the bees don't get blown away as they come back from foraging laden with heather nectar - they get straight into the hive unhindered. There is also room for the bees to cluster in the funnel to aid fanning and ventilation for honey ripening.
- When transporting the bees it is very simple to shut in the bees with a block of foam which is inserted in the funnel.

7 MOVING OR TRANSPORTING COLONIES OF BEES

Moving bees is quite a stressful time for both the beekeeper and more importantly for the bees. It should, and can, be a relatively straightforward task if planned well and all equipment is in good condition, properly assembled and other contingencies are prepared for.

The distance to move the bees does pose a few questions, all of which need to be taken into consideration. Additionally, the type of hive in use will, to a certain extent, determine the precautions that need to be taken, I am offering general advice based on BS National hives. I am aware that commercial beekeepers will probably be using Langstroth or commercial-type hives and will no doubt have alternative methods for handling many colonies in a short time. I also run a number of WBC hives, for various reasons, and yes I do move these to the moors. In fact, apart from the additional space they take up in the vehicle, they are actually far easier to move than a single-walled hive.

When moving the hives to the heather consider the best mode of transport. Many people in my view try to attempt things without thinking them through. Personally, I would never consider moving bees by trailer. Even the best double-axle trailers are not really designed for the comfortable transportation of bees, the hives get shaken and buffeted far too much. I prefer to transport my colonies in either the back of an estate car or a 4x4 pickup-type truck.

Many beekeepers use trailers - and even horse-boxes for moving their colonies. Despite both trailers here having double axles, they do not give the bees a comfortable ride. *(John Phipps)*

The boxes must be strapped together securely. Nowadays, with the advances in bee-engineering, there are some very good relatively inexpensive hive straps available that perform very well. Years ago we used to rely on bailer twine and "boy-scout" knots; invariable we had more loose bees to deal with.

If you are unsure about the security of hives when strapped together, then use a couple of straps, not at 90 degrees to each other but placed parallel to each other, that way the individual boxes cannot actually twist and let out bees. I personally dislike the various spring fastenings, knock in staples and toggle-type devices that are available, as they tend to be far too fiddly and time consuming to operate.

Give yourself plenty of time to do the moving, don't rush as that is when things go wrong. The preparation and timing of the exercise is quite important. The hives can be prepared in advance by strapping them together, placing ventilation mesh on the top and having entrance closures ready. I can block a dozen colonies in, load them up, take them to the moors and be back home in around an hour with a round trip of about 20 miles.

Carefully consider the time of day you plan to move the hives - there are many advantages and disadvantages for choosing either evening or early morning. It really does come down to a matter of personal preference and how it fits in with normal domestic activities.

The main advantage of moving the hives in early morning is that you are not chasing loss of daylight, but the down side is that you are against rising temperatures. If you do adopt the early morning approach the latest time you should aim to complete the move is 7 am, which may mean a 5 am start. Morning does not give so many options if things go wrong, especially if bees are escaping. I often load up and take hives to the moors by getting up early and dropping them off on my way to work.

Late evening is favoured by many people, but the main problem is due to fading light. If things go wrong, sorting problems out in the dark and summoning up assistance can also be more difficult as people may have gone to bed. However, if bees are escaping in large numbers it is relatively simple to leave the hive, the bees will re-enter, then you can take action the following morning and complete the task (not possible if day time moves are chosen).

Prepare for things to go wrong and make suitable contingencies. A really useful item to have is a roll of 50 mm 'gaffer' or 'duct' tape, any gaps through which bees are escaping can then be quickly and easily taped over before the bees become a major problem. It is also useful to have the roll at hand by keeping it placed over the car gear change lever, then you are not searching in the car footwells for it.

It is rather unlikely that your smoker will be ready and lit, so it is of little use if bees are escaping. There are a variety of spray-type deterrents e.g. fabispray and apifuge which are really useful as these can be used very quickly, so too can a normal plant-type water mist sprayer.

It is also good advice to wear your bee-suit as a precaution and have gloves available. I must confess I do not usually wear my suit when moving bees – but it is always available in the car, just in case.

If bees do escape during transport there is no need to worry. Although it can be a little un-nerving if you are not used to it, the bees usually pose no hazard as they will fly to the glass in the car and really not bother you at all. The only time I have been stung with bees loose in the car is when I have sat on one, thus trapping the bee.

For those who have moved hives and not yet managed to dislodge boxes or drop a hive when moving, you have done well, but don't get over-confident as it will certainly occur at some stage.

Hives in a vehicle should be positioned so that the frames are parallel with the road, this ensures they do not swing excessively during braking and acceleration.

Earlier I mentioned giving one empty super for every box of brood as a useful 'rule of thumb' method prior to moving. Provided you have done this, then full-top mesh travelling screens are not required for a trip of up to 50 miles. The porter bee escape hole in the glass quilt, blocked off with a section of varroa mesh is suitable enough for ventilation.

Just a small caution regarding the supers – if thin super foundation is fitted this can often drop out of the frames during transportation and not realised until the first inspection on the moors when a real mess of wild comb will be found. Drawn-out comb is much better. If you do use thin super foundation, then add it when the bees have been placed on site.

Too much top ventilation can actually make things worse and agitate the bees too much. I take away the hive roofs and transport them off the hive - but the bees are kept in the dark by putting hessian sacks over the top of the glass quilts.

On arrival at the moor position the colonies directly on level ground with a piece of thin plywood (or similar) in front of each hive to prevent any grass or bracken growing up and blocking the entrance. It is best to leave the bees to settle down for a while after being shaken up during the move, a minimum of 5-10 mins should be allowed, before removing the foam entrance blocks. I also count the foam pieces to make sure I have the same number as the number of hives moved – as a final check to make sure they are all removed. Leave the straps on so that all the hive parts are protected from being dislodged by wind or sheep.

Colonies left on a trailer.

8 TIME ON THE MOOR

Tutoring a new beekeeper on the art of heather honey production.

Let us assume the bees were taken to the moor in early August and the weather has been kind and the heather is lasting well - we have to be optimistic! There is always the temptation to add more supers, if the weather is looking favourable, but caution is certainly required as adding more supers can be counter-productive.

As heather honey contains a higher moisture content than flower honey it can ferment easier if stored in the unripe condition. Therefore it is far better to get one or two supers with fully capped combs than three or four that are only half-capped.

Uncapped honey.

An alternative approach to maximise the crop is to move the sealed frames to the outside and put unsealed or empty ones to the middle. This action encourages the bees to continue working in a pyramid structure. At this stage, and if the weather remains fine, then a few outer sealed frames can be removed and more unsealed frames can be added in the centre, this action is preferable to adding a full super.

Prolonged poor cold and/or wet weather can occur whilst on the moor and starvation can occur very quickly. If the weather has been poor then check the colonies food stores. The large hives are the most vulnerable due to the food demand, they can starve to death in a matter of a couple of days, feeding on the moor is not unheard of.

Also, be prepared in a poor year to have to remove supers that are not being worked, in order to maximise sealed honey.

9 REMOVING HONEY

Assuming the weather has been kind and the nectar has yielded well resulting in a good crop of surplus honey, then the next stage it to get some removed to be able to process it. Also there is likely the need to remove some of the surplus honey in order to be able to reduce the height and weight of the hives for transporting them back home.

The bees have started to cap the honey with the characteristic white wax. As soon as supers are full of sealed honey they should be removed as this will mean less heavy hives to deal with when transporting them back from the moors. (*John Phipps*)

Removing and transporting large hives full of honey is not easy, due to the weight, so it is likely that the supers will need to be removed. At this time of year and at the end of a honey flow, robbing can very easily be initiated so the best method is to use clearer boards - we use glass quilts with one centre porter bee escape.

Getting ready for returning from the moor

Brushing, shaking or blowing off bees to clear frames is to be avoided at this time of year, as this is more likely to encourage robbing.

We never remove any honey from the bottom brood box - only from the supers.

Rapid clearing with clearer boards can be ensured if there is a clear space below and above the porter bee escape. In a National hive having 11 frames the porter escape is naturally in the centre and it is best to remove the centre frame, then clearing in 24 hours will be likely. Remember if there is any brood in the frames to be removed these will not be cleared as the nurse bees will never leave the brood.

It is also not a good idea to remove every super as expansion space is needed for the trip home.

10 PROCESSING THE CROP

The next section has an important assumption — i.e. that there has been some good weather and there is actually some heather honey to be removed!

There are a number of options for processing heather honey. What you choose is largely dependant on the scale of your operation and therefore the number of combs to be processed.

The decision whether to process the surplus honey as comb or liquid honey has also to be made. I am assuming a mixture of each is likely to be required.

By far the easiest and simplest method is to cut out the comb into suitable-sized pieces and place them in proprietary 'cut comb' containers. These types of containers are very well priced and display comb honey very effectively. Thin super foundation, which is also unwired, is specifically produced for this purpose. If you have wire reinforced foundation then there must never be any wire found in the cut comb. I did once judge at a show where I found a 25 mm long length of wire in a piece of cut comb and obviously this had to be disqualified immediately. Avoid the inclusion of pollen cells in the cut comb as customers will see this as a contaminant. For selling purposes it is much more attractive to see a well-cut piece of even coloured fully-sealed comb. For 'give-aways' then the offcuts are suitable even if they are not fully capped.

One of the simplest ways of dealing with heather honey is to sell it as 'cut comb'. The comb cutter, together with the appropriate sized containers, makes easy work for the beekeeper and allows the consumer to have the honey in its most perfect form.

Ling heather honey has a thick gel-like consistency (called thixotropic) and does not run easily like other honey types. This can cause issues when trying to fine filter and process the crop.

For liquid honey production the more widely used or traditional method is to cut out the comb from the frame and press it out, usually through some form of strainer cloth such as butter muslin, hessian, or nylon mesh. The whole process of pressing heather honey is a very messy and a time consuming process.

I would recommend the nylon type of filter cloth as there is much less chance of any stray fibres making their way into the honey. The nylon cloth is also much easier to launder and dries much faster, it also lasts for a considerable time.

Traditional heather honey presses are difficult to locate, take up a reasonable amount of storage space and are often quite expensive items, however, many associations have ones on loan which are available to members.

Heather press

For small amounts of heather honey, such as a single super or so, consider scraping the honey and comb back to the foundation, this has an advantage in that the foundation is preserved for the following season. The scrapings can go into a vacuum type filter unit, the vacuum soon pulls the honey through as long as a little-and-often amount is placed in the strainer.

Vacuum strainer tank

In order to process the honey from the frames and into jars it is more efficient to carry out the process if the combs are first warmed. Careful consideration of temperatures is required and much of the information is available from many differing sources. However, to keep things very simple, the temperature combs experience in the hive is sufficient to process the honey. That temperature would be 36 degrees C, above this temperature the combs are likely to collapse.

Warming cabinets can be complicated and expensive units. Many small scale beekeepers do not have the luxury of dedicated warming cabinets. However a really simple warming device one can be made up from a plywood box, the size of a brood box (or even a brood box). Mount two batten angled lamp holders on some timber frame and use filament lamps (halogen of about 60 Watt equivalent) and a automatic digital temperature controller. Good quality accurate digital temperature controllers can be obtained from internet auction sites for less than £20.

A digital temperature controller prevents the honey from being overheated and spoilt.

The temperature controller comes with a flying lead and a sensor that is placed at the top of the stack of supers and accurately records the temperature in the unit and switches on/off the lamps to maintain the set temperature. The set temperature and the on/off differential is easily adjusted.

I place an upturned metal covered roof on the floor, (to catch any drips of honey); add the lower box with the lamps inside; then, on top of this place a wire queen excluder, (this helps to even out the temperature); next, 2-3 supers; and then a crown board added on the top. The supers can then easily be warmed overnight, without risk of overheating, ready for processing the following day. Further improvements can be made by adding insulation etc.

We supply most of our heather honey in jars to our customers with only a few samples produced as comb honey.

Our Eskdale heather honey

We are very fortunate in having a heather honey "loosener" which was one of the very first to be imported into the country over forty years ago. The unit enables us to spin the honey out of the combs using a normal extractor but the frames need to orientated in the tangential manner.

The loosener is quite a simple concept. It has a series of small 2 mm diameter nylon, blunt ended needles that when the operating handle is moved a number of times, the needles agitate the honey in each cell a few times and make the honey temporarily liquid, which enables it to be extracted.

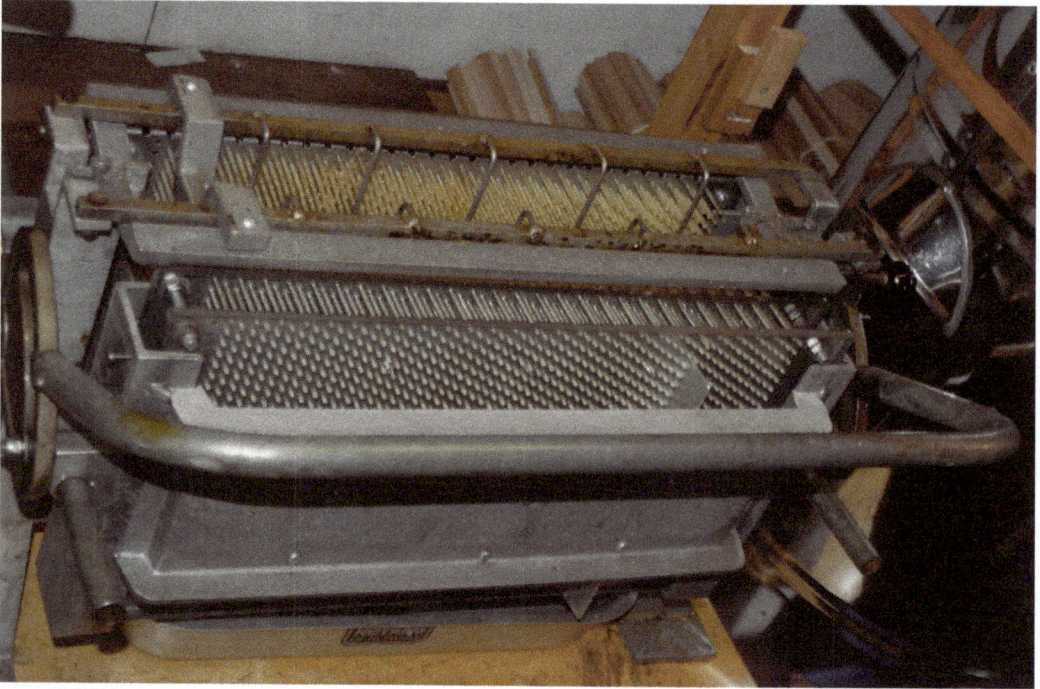

Heather honey loosener - extremely useful when large amounts of honey need processing.

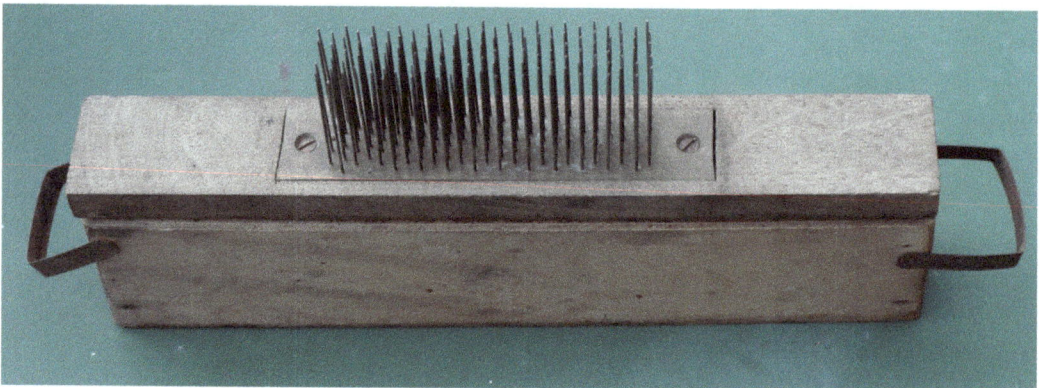

Beekeepers with only small amounts of heather which they want to extract in 'liquid' form, might be able to find a simple heather loosener at a beekeeping sale or auction. *(John Phipps)*

During storage the honey will revert back to its thixotropic state within a reasonable amount of time, but not as quickly as honey pressed out in the traditional manner.

As long as the frames are put in the extractor the correct way round, the honey is easily spun from the frames. The direction of travel of the frames in a tangential extractor is important as the bottom bar needs to face the direction of travel. This is due to the cells having a slight angle towards the foundation, and then the honey is spun from the cells. (If cells were horizontal the nectar would run out of the frames when in the hive).

Tangential extractor screens

The loosener, although initially quite expensive is a brilliant piece of equipment. The major benefit is that the drawn out combs are preserved for the following season. This gives us a major time and cost reduction from having to wax up new frames with foundation the following, usually busy spring period. The additional spring and summer honey crop is also improved as the foundation is already drawn out ready for making the most of the nectar flows.

For exhibition purposes getting really good samples of heather is quite difficult and there are actually not that many judges that regularly come across fine quality heather honey. Heather honey is thick (thixotropic) and difficult to fine filter, therefore it is difficult to remove wax pieces when preparing it for showing. As a sample for the show bench, the best method is to press out the honey through fine filter cloth.

11 AFTER THE CROP

Leaving only heather on the hives as a sole winter food is never a good idea. There is a tendency (especially in lean years) to take out heather from the lower brood chamber. I don't practise this as any honey in the brood box is for the bees own use. Too much heather honey in the winter diet, however, often results in dysentery around Christmas time. It is much better for the bees if a balance of food is available during the winter period. Hopefully, a late income of nectar from ivy has occurred and additional feeding, with a thick syrup in the autumn and fondant from beginning of November, assists in providing a more balance supply of food over the winter.

Nectar from ivy will be stored by the bees and used first during the winter. Together with the syrup that is fed, it will delay bees using the pollen-rich heather honey until much later when it will be especially usefully when they begin rearing brood. (NB If we find yellow-banded bees in our colonies, we keep them for that season and then re-queen or unite them with native colonies). (John Phipps)

There is no need to rush the process of getting the bees back, removing the honey, carrying out varroa treatments and sorting them out for winter.

All my hives are reduced to a single BS national brood box for the winter period. It can take a while for large colonies with the brood in the half brood box to hatch, enabling the reduction to take place, patience is required by the beekeeper. I often hear advice of adding a super for winter, sometimes under the brood, sometimes over – there is just too much conflicting advice. I find that bees in winter will not pass over a bee space from a brood box to a super and it is just far too much space for them to regulate the hive temperature. I use BS National brood boxes placed on our version of a WBC flat (unvented) floor, with two WBC lifts surrounding them, and the space above the glass quilt filled with layers of insulation (usually carpet and hessian).

Mouse guards are fitted after the last chance of ivy pollen gathering. I use long-slot galvanised queen excluders cut into 50 mm wide strips and placed on the floor over the slot in our design of floor - making sure no mice have already entered the hive.

I am a firm believer that the ambient temperature for varroa control with thymol-based products is not really an issue as long as the hive is reduced to a single brood box prior to feeding for winter and then applying the varroa treatment directly at the brood area, The heat of the small amount of brood is sufficient for the thymol to function. Our thymol treatment is applied into October, as it also knocks down more adult mites.

Some years ago we started to apply fondant feed from the beginning of November and have continued to do this since. At one of our apiaries it can be difficult to get into it if there has been significant rainfall, as the access to the apiary is over a beck which can be difficult, so applying fondant early is good insurance. Fondant is warmed slightly and 2/3rds placed in honey jars and topped off with honey (the skimmed off, frothy honey which has been removed from storage tins after extracting) so the feed is a mixture of fondant and honey. The sealed jars are kept until they are required and simply placed over the glass quilt oval porter bee escape hole. A single piece of carpet with a 30 - 50 mm diameter hole is placed over the glass quilt first as it then assists in sealing off the oval hole to that of a single jar top. Also, consider - why have the fondant stored in the bee shed, when it may as well be placed where it is needed - on top of the hive. Check the hives at a frequency of every two weeks and replenish any jars of fondant that have been used during the Winter period.

As I have applied a thymol-based varroa treatment much later in the year than is normally advised (when brood rearing is much reduced anyway), together with a solid floor and a twin-walled hive construction, I do not partake in further varroa treatments, until the situation is reviewed in the following spring.

I also wonder if anyone can come up with a better solution for keeping hive roofs in place, the good old-fashioned house brick seems to be so widely used - and we are in a three-brick windy area.

Keeping an eye on the apiary and dealing with any issues as they arise is all that needs to be done for the rest of the winter period whilst watching patiently for the first pollen to become available.

In my view spring time should be the start of the beekeeping year, many other people will advise us that the beekeeping year starts at other times, for various reasons. To me the availability of the first pollen is of crucial importance, normally this comes with the pussy willow catkins and small clumps of snowdrops. At this time a floor change is carried out, mouse guards are removed - as they tend to strip off the vital incoming pollen from the bees, and a stimulative 1:1 feed of syrup in contact feeders (old honey jars) is given to the colonies to make sure the queens start to lay.

There is no rush to carry out spring inspections. In the early spring there is very little that can be done if you find something amiss with a hive. For example, if it is queenless, or a drone layer – what can you do? Any remaining bees are going to be elderly bees and probably not worth nurturing anyway. This is the main reason for not wishing to commence first inspections until after Easter. After all, when the season does start in earnest, it is such a busy time anyway, why become busy before you really need to be?

The stimulative feeding started when pollen became available should set in motion the production of eggs that will become foraging bees after six weeks - which should coincide with the OSR nectar flow. This, in turn, will once again, be the commencement of the work leading up to its August climax.

You often have to wonder why we involve ourselves in such a lot of work - but the honey we harvest is such a highly-prized product which is well worth the effort; to see lots of thick, high aroma, wonderfully strong tasting honey in jars and buckets ready to go out for sale is sheer delight.

By concentrating on heather honey production as outlined in this short book, you will easily see that the production of strong healthy bees brings rewards for the other spring and summer crops of honey as well as producing lots of new colonies with plenty of drawn comb. Most colonies will build up to be production colonies and any excess colonies can be sold off to set up new beekeepers in the craft.

SUMMARY

Key aspects – maximising the crop

Breed for foraging bees

Have foraging bees available and ready for beginning of August

Prepare transport and secure hives

Plan contingency equipment

Move colonies onto moor beginning of August

From egg to foraging bee = 6 weeks

Eggs need to be produced early to mid-June

Queen excluders are not necessary

Robbing on moors is very likely

Drawn comb is vital

Lightning Source UK Ltd.
Milton Keynes UK
UKHW051008150223
417021UK00004B/30